Space Exploration Activity Book

Author	Mary Jo Keller
Editor	Kathy Rogers
Illustrations	Barb Lorseyedi
Cover Design	Imaginings

METRIC CONVERSION CHART

Refer to this chart when metric conversions are not found within the activity.

¼ tsp	=	1 ml	350° F	=	180° C
½ tsp	=	2 ml	375° F	=	190° C
1 tsp	=	5 ml	400° F	=	200° C
1 Tbsp	=	15 ml	425° F	=	216° C
¼ cup	=	60 ml	1 inch	=	2.54 cm
⅓ cup	=	80 ml	1 foot	=	30 cm
½ cup	=	125 ml	1 yard	=	91 cm
1 cup	=	250 ml	1 mile	=	1.6 km
1 oz.	=	28 g			
1 lb.	=	.45 kg			

Reproducible for classroom use only.

Not for use by an entire school or school system.

EP122 • ©1996, 2003 Edupress, Inc.™ • P.O. Box 883 • Dana Point, CA 92629

www.edupressinc.com

ISBN 1-56472-122-1

Printed in USA

Table of Contents

 © Edupress EP122

Literature List

- ***Voyager to the Planets***
by Necia H. Apfel;
Houghton 1991. (3-6)
This account follows the spacecraft from its 1977 launch to Jupiter flybys.

- ***Astronaut Training***
by Ann Armbruster & Elizabeth Taylor;
Watts 1991. (4-6)
An inside, full-color look at NASA and astronaut training.

- ***Space, Stars, Planets and Spacecraft***
by Sue Becklake;
Dorling LB 1991. (4-7)
An explanation of how complex machines work in space and their contribution to life on Earth.

- ***Artificial Satellites; Helpers in Space***
by Jeanne Bendick;
Millbrook LB 1991. (2-4)
The purposes and composition of space satellites are explained with many illustrations.

- ***Living in Space***
by Don Berliner;
Lerner 1993. (4-6)
The living conditions of travelers in space are covered.

- ***Our Future in Space***
by Don Berliner;
Lerner LB 1991. (4-7)
Answers such questions as how do astronauts bathe and can we build a permanent community in the sky.

- ***The Day We Walked on the Moon: A Photo History of Space Exploration***
by George Sullivan;
Scholastic 1990. (5-8)
The history of U.S. space exploration, showing the accomplishments of both the United States and the Soviet Union.

- ***To Space and Back***
by Sally Ride & Susan Okie;
Lothrop 1986. (3-7)
The first American woman in space describes her experiences aboard the shuttle.

- ***Operation Rescue: Satellite Maintenance and Repair***
by D. J. Herda;
Watts LB 1990. (3-6)
The retrieval and repair of orbiting satellites and a history of the U.S. space program.

- ***Apollo 11***
by R. Conrad Stein;
Childrens LB 1992. (3-5)
The story of the *Apollo 11* flight to the moon in simple text and many pictures.

- ***Pioneering Space***
by Sandra Markle;
Macmillan 1992. (5-8)
Presents basic information on space travel and how people will live in space.

- ***Space Probes to the Planets***
by Fay Robinson;
Whitman LB 1993. (2-4)
Stunning color photos and clear text tell the story of the space probe.

- ***An Album of Voyager***
by Maury Solomon;
Watts LB 1990. (4-7)
A collection of captioned photos of the journeys of the two *Voyagers* to various planets.

- ***The Picture World of Rockets and Satellites***
by Norman Barrett;
Watts LB 1990. (K-4)
This book introduces satellites and rocket types and includes a look at the Hubble space telescope.

© Edupress EP122

Glossary

astro—used to mean space in words such as astronaut and astronautics.

aeronautics—the science of designing, making, and operating aircraft.

astronaut—a person trained to make rocket flights in outer space.

astronautics —the science of space flight.

booster—the rocket system that launches a spacecraft.

capsule—module on the rocket that holds the astronauts or instruments. It is usually recovered after flight.

cosmonaut—a Russian space pilot.

countdown—the schedule of all the operations to check the equipment and systems just before the firing of a rocket.

escape velocity—the speed a spacecraft must reach to coast away from the pull of gravity.

extravehicular activity—a task performed by an astronaut outside a spacecraft in space.

launch pad—the platform from which a rocket is blasted into space.

launch vehicle—a rocket or combination of rockets used to launch a satellite or other spacecraft into orbit.

launch window—the time during a mission when the conditions are right for launching a rocket.

heat shield—spacecraft covering that protects the astronauts during reentry.

microgravity—the only gravity on a spacecraft. It equals 1-millionth of the gravitational force felt on Earth.

module—a single section of a spacecraft that can be separated from the other sections.

NASA—the National Aeronautics and Space Administration.

orbit—the path taken by a satellite around a body in space.

propellant—the substance burned in a rocket to produce thrust.

reentry—the part of the flight when a returning spacecraft begins to descend through the atmosphere.

rendezvous—a space maneuver in which two or more spacecraft meet.

revolution—one complete cycle of a satellite in its orbit.

spacecraft—a man-made object that travels through space.

satellite—natural or artificial body that revolves around a larger body, such as a planet.

stage—one of two or more rockets combined to form a launch vehicle.

thrust—the push given to a rocket by its engines.

trajectory—the curved path of an object hurtling through space.

weightlessness—the conditions of objects in a spacecraft when the gravitational force is counterbalanced by the ship's motion.

© Edupress EP122

Rockets

Information

In the late 1600s the English scientist Sir Isaac Newton studied gravity and motion. Newton's Third Law of Motion stated that for every action there is an equal and opposite reaction. "Action" and "reaction" are forces that act in opposite directions. When a rocket ignites, the fuel burns and makes huge amounts of hot gases that expand and shoot out the back of the rocket. As the gases thrust downward, the rocket is pushed upward! In other words, the *action* of the gases pushing downward is accompanied by the *reaction* that pushes the engine upward.

Rockets fall into two main categories—liquid fuel or solid fuel. A liquid fuel rocket has two tanks, one containing fuel and the other containing liquid oxygen. The two fuels are pumped into a combustion chamber where they are ignited and burned. The fuel in a solid fuel rocket looks and feels like a pink pencil eraser! It is shaped into a tube with a small hole down the center. An *igniter* starts the fuel burning outward from the central hole.

Project

Demonstrate Newton's Third Law of Motion.

Materials

- Two skateboards
- Helmets

Directions

1. Sit on one skateboard and have a friend sit on the second skateboard facing you.
2. Push against your friend. What do you suppose will happen to your skateboard when you do? What do you suppose will happen to your friend's skateboard? The *action* of you pushing will cause the other skateboard to move. Notice that the *reaction* to your push forces your skateboard to move in the opposite direction. This is Newton's Third Law of Motion in action. Can you think of any other ways you could demonstrate this principle?

 © Edupress EP122

Rocket Pioneers

Information

About 1000 years ago, the Chinese invented the first rockets soon after they discovered how to make gunpowder.

In 1903, a Russian school teacher named Konstantin Tsiolkovsky published a paper on the use of rockets in space flight. His idea was to use liquid-propellant rockets instead of gunpowder.

Hermann Oberth, a researcher in Germany, studied the technical problems of space flight. He published a book in 1923 that described what a spaceship would be like.

In 1926, Robert Goddard, an American scientist, launched the first successful liquid propellant rocket. The rocket burned gasoline and liquid oxygen. Goddard had been fascinated with rockets since childhood.

Project

Commemorate the work of rocket pioneers with a space launch.

Materials

- Balloons (tube-shaped balloons work best)
- Tape
- Straws
- String
- Scissors

Directions

1. Divide into three teams—Team Tsiolkovsky (tsee-ol-KOV-skee), Team Oberth, and Team Goddard.
2. Cut the string into lengths about ten yards (9.14 m) long. Depending on the size of the balloons, your rockets may go even farther! Make sure each team has the same length string.
3. Thread the string through the straws. Use the tape to attach the balloon to the straw. Have one team member hold each end of the string. Have a third team member blow up the balloon and hold the end tightly closed.
5. On the signal, let go of the balloons and see which rocket goes the farthest!

© Edupress EP122

Launch Vehicles

Information

Launch vehicles are rockets used to launch spacecraft, satellites, and probes into space. A launch vehicle consists mostly of fuel tanks that hold the large amount of fuel a rocket needs to burn in order to get into orbit.

Many launch vehicles are made up of three or more sets of rockets called stages. Most of the *propulsion,* or push, that carries the spacecraft into orbit is provided by the first, or *booster,* stage. As each rocket runs out of fuel and falls away, the next stage takes over until what is left of the launch vehicle is at *escape velocity*, the speed that will take it into orbit.

The Saturn V Moon rocket is the biggest rocket ever built. It carried the first astronauts to the moon. The booster stage of this giant rocket measured 140 feet (42.7 m) high and its overall height was 363 feet (110.6 m) high! It had the *thrust,* or push, given to a rocket by its engines, of 40 jumbo jets! In comparison, the space shuttle system measures only 184 feet (56 m) in height.

Project

Visualize the massive size of different launch vehicles.

Materials

- Tape measure
- School bus
- Paper and pencil

Directions

1. Measure the length of a school bus.
2. Calculate how many school busses you would have to line up in a row to equal the height of the Saturn V. How many school busses lined up equal the space shuttle system? Research other launch vehicles such as Jupiter C and Titan II and chart your results.
3. If you have a big enough area, measure out the length of the Saturn V. How long did it take you to jog from one end of this gigantic launch vehicle to the other?

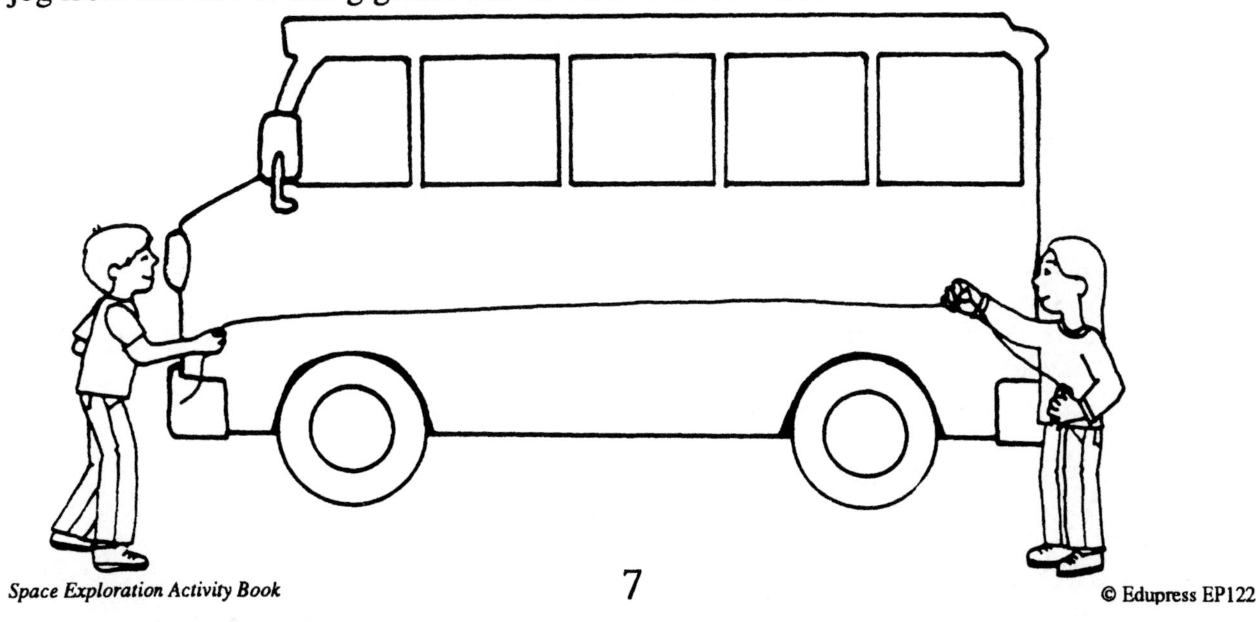

© Edupress EP122

Astronauts

Information

A person who is trained to make rocket flights in space is called an *astronaut*. This word comes from two Greek words that mean "sailor among the stars." In Russia, space travelers are called *cosmonauts*, or "sailors of the universe."

Pilot astronauts are trained to command and control a spacecraft. *Mission specialists* are trained to perform a variety of tasks: conducting various experiments, launching and recapturing satellites, and maintaining the spacecraft and its equipment.

The first person to travel in space was Yuri A. Gagarin of the former Soviet Union. On April 12, 1961, Gagarin spent one hour and 48 minutes in space. During that time, he orbited the Earth one time.

The first American in space was Alan B. Shepard, Jr. Only 23 days after Gagarin's historic flight, Shepard made a 15-minute trip into space, but did not go all the way into orbit.

Project

Research outstanding accomplishments of some famous astronauts and cosmonauts.

Directions

1. Reproduce the "They're Out of This World" pages on card stock and cut apart.
2. Conduct research to find three facts about each of the astronauts. Write the facts on the trading cards.
3. Use the trading cards to play a game of "Name that Space Traveler."

Materials

- "They're Out of This World" trading cards, following
- Card stock
- Pencils
- Resource Books
- Scissors

Something to think about!

Imagine you are in charge of naming the group of pilots and specialists your country is about to send into space. What name would you give these space travelers?

 © Edupress EP122

Valentina Tereshkova Cosmonaut

Guion S. Bluford, Jr. Astronaut

Sally K. Ride Astronaut

Alexei A. Leonov Cosmonaut

Steven McLean Astronaut

Roberta Bondar Astronaut

© Edupress EP122

Yuri A. Gagarin
Cosmonaut
Alan B
Shepard, Jr.
Astronaut
Harrison H.
Schmitt
Astronaut
John Glenn, Jr.
Astronaut
Marc Garneau
Astronaut
Neil A.
Armstrong
Astronaut

Astronaut Training

Information

It takes a year of study and training at the Johnson Space Center to become an astronaut. Candidates study subjects such as *aerodynamics* and spacecraft-tracking techniques. They learn about flight-control systems, and on-board equipment. *Flight training* takes place in jet aircraft where candidates practice maneuvers at high altitudes. Future astronauts even experience near-weightlessness. As their large training airplane begins to go into a dive, the trainees float in the padded cabin for about 30 seconds.

When an astronaut is assigned to a mission, he or she spends many hours training in a *simulator* that reproduces the conditions of space flight. Full-size models of spacecraft help astronauts practice the tasks they will perform in space—everything from entering and leaving the spacecraft to preparing a meal! Special assignments may require additional instruction such as training with a jet-powered backpack as practice for a satellite recovery mission.

Project

Practice some of the exercises that keep the astronauts in top physical condition during their training.

Materials

- Room to exercise

Directions

1. Pretend you are an astronaut. Try doing some of the exercises the astronauts do.
2. Keep an exercise log for at least one week. Record how many of the exercises you were able to do in one minute or how far you were able to stretch. How much did you improve after one week?

ASTRONAUT EXERCISES

Curl-ups—Lie on a cushioned surface, knees flexed. Have a partner hold your feet. Cross arms across your chest. Raise upper body until elbows touch the thighs, then lower yourself back down until shoulder blades touch the floor.

Sit and Reach—Sit on the floor, feet shoulder-width apart. How far forward can you reach?

Pull-ups—Hang from a horizontal bar. Raise body until chin clears the bar. Lower body to full-hang position.

Push-ups—Lie face down on a mat, hands under your shoulders, fingers straight. Legs should be straight, parallel, and slightly apart, toes supporting the feet. Straighten arms, keeping back and knees straight. Then lower body until upper arms are parallel to the floor.

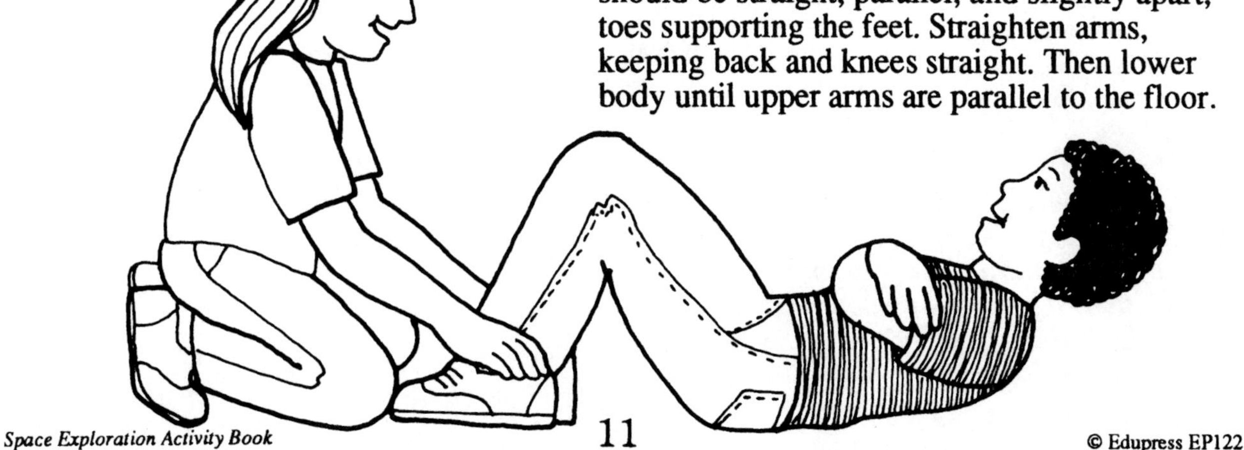

 © Edupress EP122

Countdown to Launch

Information

Space vehicles are launched from specially constructed launchpads. The equipment needed depends on the size of the vehicle. A huge spacecraft like the space shuttle requires a launch complex with several specially-designed buildings and two launchpads. The Vehicle Assembly Building where the shuttle is connected to its rocket boosters and fuel tank is 52 stories tall!

Before the launch, the spacecraft and the rocket that propels it are set up and carefully inspected. Weather conditions are also closely monitored to ensure ideal launch conditions. Engineers and technicians in the nearby control center follow a schedule to be sure that all the preparations are completed at the proper time. This schedule is called a *countdown*.

When preparations are complete, the rocket engines are fired. The rocket and the spacecraft lift off the ground and begin the journey into space.

Project

Monitor the weather to see if your community could host a rocket launch.

Directions

1. Reproduce the Weather Countdown Chart.
2. Using weather reports from the internet, newspapers, radio, or television, research the weather in your community. Record the findings on your chart. Would today be a good day for a rocket launch? Why or why not?
3. You can compare the weather in your city with that of Cape Canaveral in Florida. Record the results on a separate chart and compare. Does either place have ideal weather for a launch? Which place has more of the required weather conditions needed for a launch?

Fun to Know

The Vehicle Assembly Building is the largest building in the world. If the American flag that is painted on the side of the building were painted on the ground, a school bus could easily drive along one of the thirteen stripes!

Materials

- Pen or pencil
- Newspaper
- Newspaper, radio, internet, and/or television weather reports
- Weather Countdown Chart, following

 © Edupress EP122

Weather Countdown Chart

City Name ______________________________

Weather	***Required Conditions***	***Current Conditions***	***Meets Requirements?***
Clear skies	No fog or haze No lightning		
Wind	Under 12 miles (19.3 km) per hour		
Precipitation	No rainfall within 20 miles (32 km)		
Clouds	No clouds lower than 8,000 feet (2,438 m)		
Current Temperature	Above 41°F (5°C)		
Average Temperature—past 24 hours	Above 41°F (5°C)		

© Edupress EP122

Dressing for Space

Information

Before an astronaut can venture out into space, he must put on several layers of special clothing. The first layer is like a pair of long underwear that has water-cooling tubes running all through it. This layer keeps the astronaut at a comfortable temperature.

The space suit itself is also made up of several layers. These layers were designed to protect the wearer from the many dangers found in space such as extreme temperatures, radiation, and *micrometeorites*, or space dust. The inside layer is a pressure bladder—like a flat balloon that is filled with oxygen. Next comes a layer of plastic for strength and several layers of fireproof material and thin sheets of metal.

Early space suits were connected to the life-support system of the spacecraft by a tube called an *umbilical*. Space suits worn today have a life support system backpack built right into the upper part of the suit.

Project

Enjoy a simulation activity that takes students through the process of dressing for space.

Materials

- Activity script, following
- Clothing: tights or long underwear, pants, boots long-sleeved T-shirt, knit hat, gloves, helmet

Directions

1. Have the students stand in an area where they have plenty of room to move about.
2. Instruct the children to put on each layer of clothing as you read aloud. Speak slowly and pause to give students enough time to act out their activity. Make sure to give the children plenty of time to explore each step before moving on.

 © Edupress EP122

Imagine that you are an astronaut who has a task to perform outside the spacecraft. It is time to get into your space suit. Imagine that you are inside an airlock on the space shuttle.

1. Let's begin by putting on the first layer of your space suit. It is like pulling on a pair of long underwear. It has tubes running all through it so it isn't as easy to get into as long underwear! You have to put your legs in first, one at a time, then wiggle the suit high enough to get your arms into the openings. Now fasten this layer closed.
2. Next you must climb into your space trousers. They are very thick and bulky. The boots are connected to the trousers. Wiggle your feet to get them comfortable inside all the protective layers.
3. The torso of the suit is the next part you need to put on. Torso means upper body, so you know you have to put your arms into this part of your space suit. You need to squat down and reach up into the torso of the suit. Pull it down over your head. There are thumb loops on the undergarment. Make sure the loops are over your thumbs!
4. Make sure you have connected the tubes in your layers to the life support system. Now pull the flap on the torso down over the trousers.
5. Put on the communications carrier which is like a hood with a headset built in. Adjust the headphones so they fit snugly over your ears. Test the microphone. Adjust the oxygen flow in the suit—the controls are right on the front of the suit.
6. Put on the gloves and wiggle your fingers inside the thick gloves. Snap the closures around the wrist, and lock the connecting rings by sliding the tab into the locked position.
7. Put on the helmet. Make sure your helmet is resting on the right spot of the torso because now you have to twist it to lock it onto your space suit.
8. You are ready to climb out of the hatch of the airlock into the cargo bay of the shuttle. Attach your life-line to the safety wire that runs along the side of the bay.
9. Pretend that you are floating in space inside your thick space cocoon!

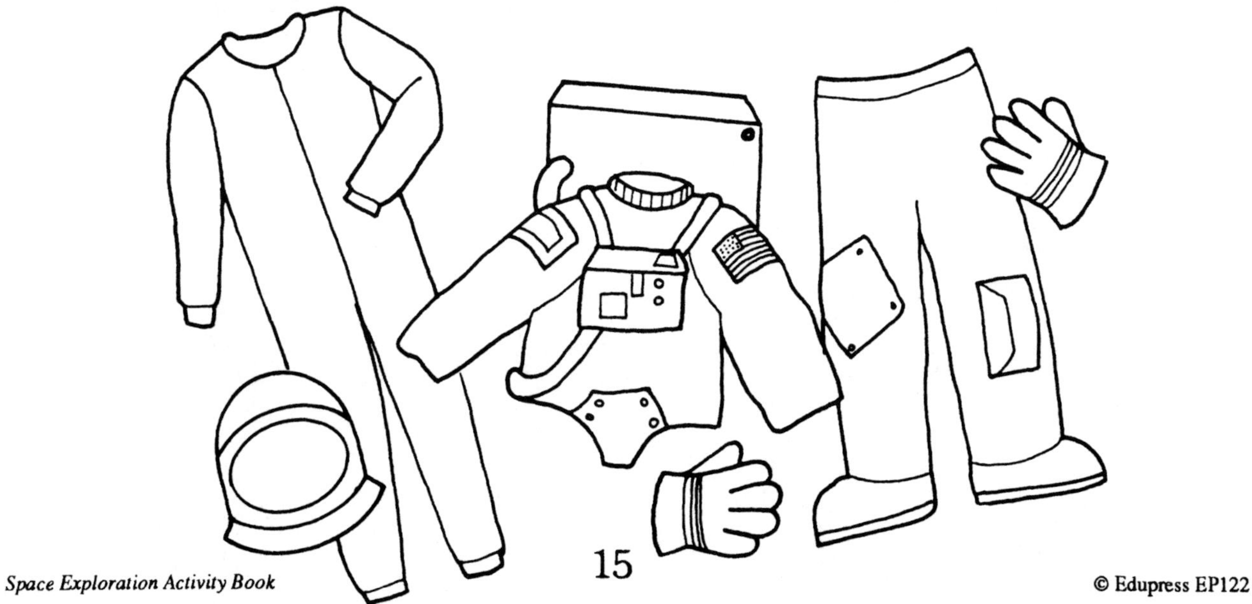

© Edupress EP122

Eating in Space

Information

Eating in space is not easy! You can't pour milk into a glass and food won't stay on a plate! Much of what the astronauts ate on early space missions was food that had been *freeze-dried.* The food was stored in tube-shaped plastic bags. When it was time to eat, the astronaut would add hot water to one end of the bag. After mixing the food and water together for a few minutes, the food was ready to be eaten. The astronaut would have to cut off the end of the bag and squeeze the contents into his mouth!

The food served to modern astronauts is much more appetizing. Shuttle astronauts use dishes with special covers and magnetized silverware and dine on meals such as shrimp cocktail, steak, and chocolate pudding.

Food in space may not have much taste. In a weightless environment astronauts often feel cold-like symptoms such as a runny nose and that makes smelling and tasting more difficult.

Project

Choose one or all of the activities and learn about eating in space.

Materials

- Refer to each activity

EATING IN SPACE

Can you eat in space if there is no gravity to carry your food from your mouth to your stomach?

Materials

- Paper cup • Water • Straw
- Bench or two chairs pushed together

1. Fill the cup with water, put the straw in it, and place it on the floor at the end of the bench.
2. Lie on the bench with your head hanging down so that your mouth is lower than your stomach. Can you drink the water? Do you think you could still eat and drink in a near weightless environment? What do you think would happen if you tried to eat a cracker while standing on your head? Could you do the same trick in space?

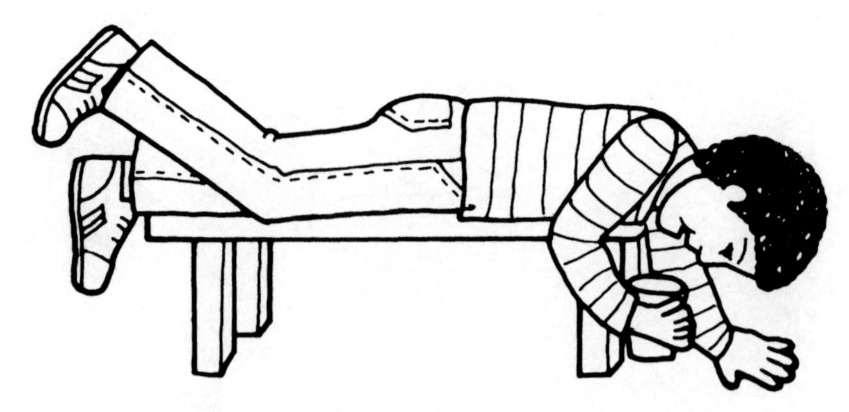

© Edupress EP122

SPACE FOOD COLLAGE

Materials
- Poster board
- Scissors
- Magazines
- Product labels
- Glue

Use pictures from magazines or actual product packages to make a poster display showing different foods that could be, or actually were, eaten in space. Freeze-dried coffee and powdered drink mix labels will get your collage started.

MENU PLANNING

Here are some guidelines for planning space meals. If a food is going to be eaten with utensils, it must stick to the fork or spoon. No foods that make crumbs! Meals must be be nutritionally balanced and high in potassium and calcium. Hot or cold water are the only additional ingredients available. No leftovers allowed!

Materials
- Paper
- Pencil

1. Plan a menu for one day's meals on the space shuttle. Each meal should be about 1,000 calories, nutritiously balanced, and as delicious as possible!
2. Use the guidelines above and the nutritional information on food packages to help you plan your menus.

SUPERMARKET SCAVENGE

You can find many of the foods developed for astronauts to eat in space at the supermarket! Single serving canned pudding is only one example. It does not need to be refrigerated, sticks to a spoon, there is no preparation, and it's tasty!

Materials
- Paper
- Pencil

Have fun shopping in your supermarket for food items you think an astronaut could easily eat in space. Make a list of your finds.

TASTING PARTY

Eating out of a zip closure bag is similar to eating from the spoon-bowl packages the astronauts' food comes packaged in!

Materials
- Very warm and cold water
- Small zip closure plastic bags
- Powdered orange drink
- Instant mashed potato flakes
- Spoons
- Straws

1. Place single serving portions of drink mix and instant potatoes in small plastic bags.
2. With adult supervision, add the warm water to the potatoes. Mix together for a few moments, then taste.
3. Repeat with the drink mix and cold water.

© Edupress EP122

Orbiting the Earth

Information

Once a rocket has carried a satellite into space it must begin to circle the Earth at the correct speed so that it does not fall right back down to Earth! A *guidance system* turns the rocket so that it is pointing in the same direction as the Earth's rotation.

Once a satellite is in orbit, no fuel is required to keep it there. Nothing "holds up" the orbiting satellite. It stays in orbit because there is a balance between the satellite's forward motion and the downward motion caused by the force of gravity. The combination of the satellite's sideways speed and its falling motion gives it a *trajectory,* or path, that matches the Earth's curved surface.

A spacecraft in a typical circular orbit about 200 miles (321.7 km) above the surface of the Earth travels at 17,000 miles (27,400 km) per hour! At that speed, it only takes about 90 minutes for the satellite to circle the Earth!

Project

Experience the pull of gravity on a satellite orbiting the Earth.

Materials

- Vertical pole mounted securely in the ground, such as the kind used for a basketball hoop
- Rope, about six feet (1.8 m) long

Directions

1. Tie one end of the rope to the pole. Do not tie it too tightly as it needs to slip around the pole.
2. Stand with your shoulder towards the pole and the rope extended out to its full length. Pretend you are a satellite and the pole is the Earth. As you start to move forward, walking around the pole, you can feel the force of the "Earth's gravity" pulling you into a circular orbit.

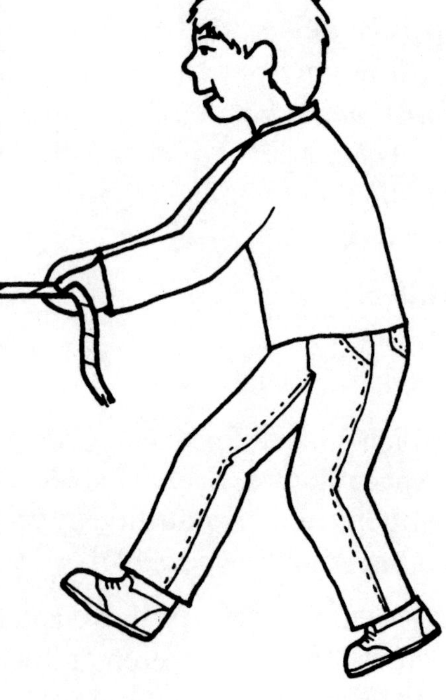

© Edupress EP122

Working in Space

Information

Most of the tasks an astronaut has to perform are carried out inside the spacecraft. But sometimes an astronaut has to go outside into space to complete a particular task. The spacecraft may need a routine inspection or repair. Perhaps the film packs are due to be changed. On a shuttle flight, they might be launching or repairing a satellite. When an astronaut leaves the spacecraft, it is called *extravehicular activity,* or EVA, for short.

Astronauts wear protective space suits when they leave the spacecraft. The astronauts who explored the moon wore the first self-contained space suits. They carried out experiments and collected samples while wearing bulky backpacks that contained their life-support equipment. On a satellite recovery mission, astronauts put on jet-propelled backpacks called MMU's, or manned maneuvering units. By firing small thruster jets, the astronauts can move through space without being attached to the spacecraft!

Project

Practice collecting "samples" wearing bulky space gloves.

Materials

- Light cotton gloves
- Rubber kitchen gloves
- Heavy workman's gloves
- Wide-mouth plastic cups
- Variety of items such as buttons, coins, paper clips, screws, etc.
- Gardening gloves
- Barbecue tongs
- Paper lunch bags

Directions

1. Place the cups on the ground. Spread out the assorted "samples" on a table.
2. Astronauts' gloves are made of several layers of protective materials. Put on the cotton gloves first, then the rubber gloves to simulate the pressure bladder layer. Next, put on the plastic gardening gloves for strength, and finally the workman's gloves for protection.
3. Try to pick up the items from the table and put them into the bag. Then try to use the barbecue tongs to pick up the cups from the ground. Can you stack the cups? Practice and see!

Fun To Know!

Each astronaut has a pair of gloves custom made for him using laser scans of his hands. The finished gloves have rib-like ridges on the finger joints and a complex pattern of diamonds and squares on the palms.

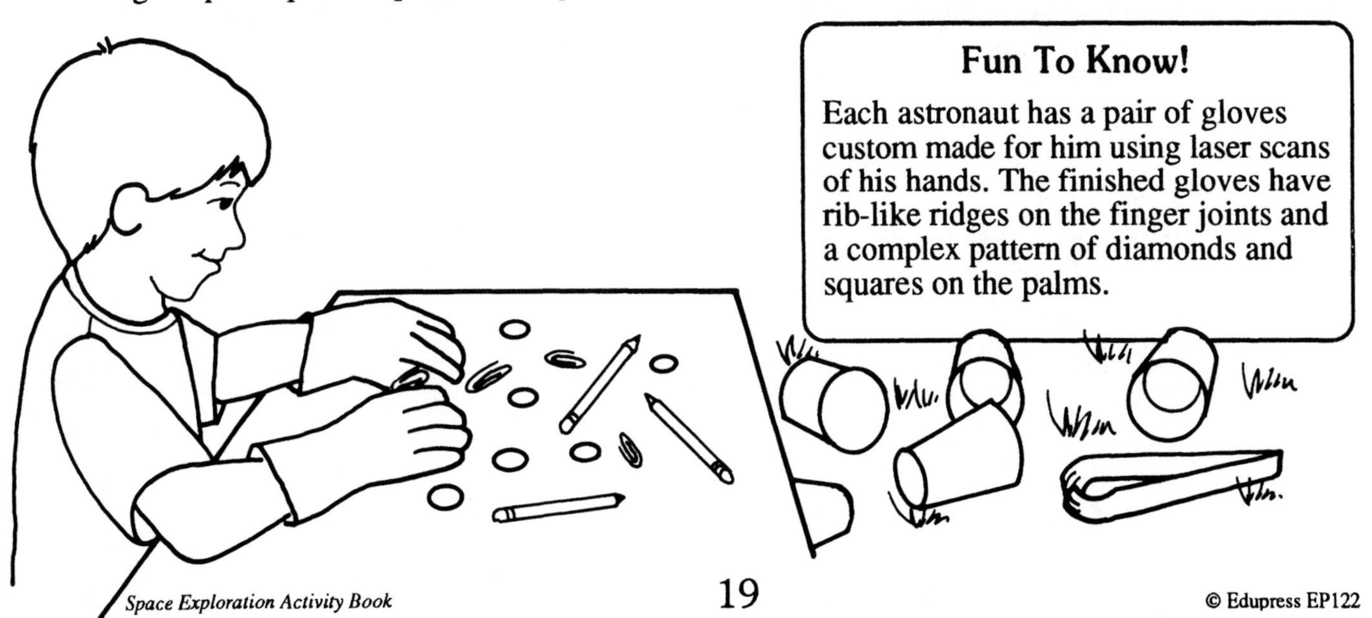

© Edupress EP122

Living in Space

Information

A day in space might include launching or repairing a satellite, collecting data on solar flares, or studying sources of pollution on Earth. As busy as they are, astronauts still must make time in their daily routine to exercise, shower, eat, sleep, play, and relax!

Taking a shower on a spacecraft requires a special folding shower stall that has a vacuum system in it to suck away the used water. The toilet on board, called the waste collector, resembles the ones found on airplanes.

Exercise is an important part of an astronaut's routine because weightlessness weakens the body. Astronauts must exercise at least 30 minutes a day. These exercises might include riding a stationary bike or working out on a device similar to a treadmill.

Sleeping compartments hold sleeping bag-like enclosures that allow astronauts to sleep without floating and without having to be strapped down.

Project

Make journal entries to describe an astronaut's journey in space.

Materials

- Journal page, following
- Resource books
- Pen or pencil
- Stapler

Directions

1. Reproduce the journal template. Fold on the dotted lines. Staple several together to make a journal.
2. Do some additional reading on your own about living in space. See the Literature List (page 3) for some suggestions.
3. Keep a journal that starts from count-down through landing. Include descriptions of what your "home in space" looks like, and also what you do throughout the day and how living in space makes you feel.

© Edupress EP122

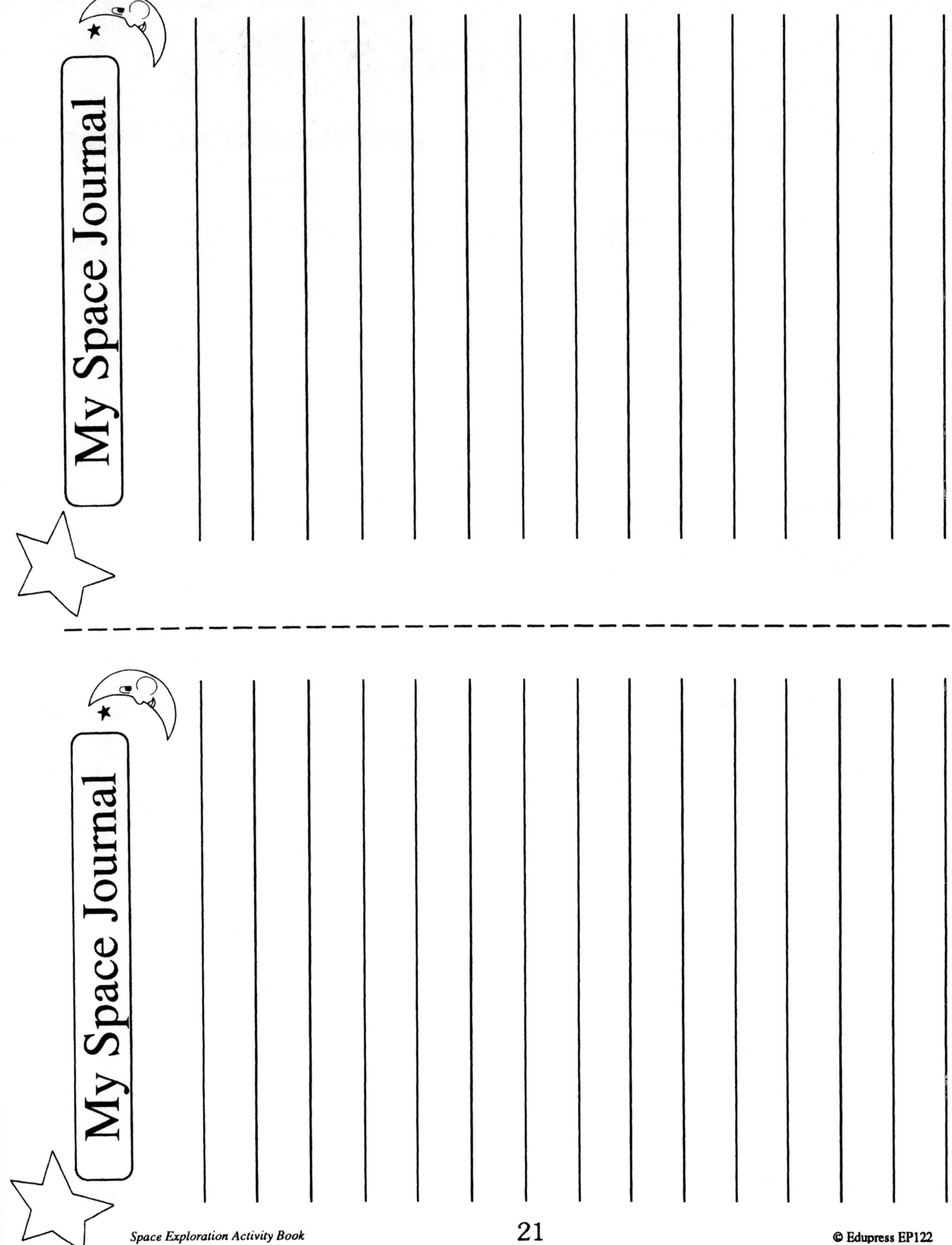

 © Edupress EP122

Return to Earth

Information

The trip back to Earth for a spacecraft and its crew involves overcoming problems that are just the opposite of getting into space! Instead of gaining speed, the spacecraft must lose speed. As the spacecraft slows down, it "rubs against" the surrounding air producing great amounts of heat. This rubbing of one surface against the other is called *friction*. The harder and faster the surfaces rub, the more heat is generated.

When the space shuttle orbiter enters the Earth's atmosphere, it is traveling almost 17,000 miles (27,400 km) per hour! The temperature of its wings can exceed 2700°F (1482.22°C). The orbiter needs protection from this intense heat. Its *thermal protection system* consists of more than 25,000 ceramic tiles that are adhered to its body

Project

Learn about friction and thermal protection systems through a variety of activities.

Materials

- See individual activities

Insulation Tests

Pretend you are a space engineer choosing a thermal protection material for a spacecraft.

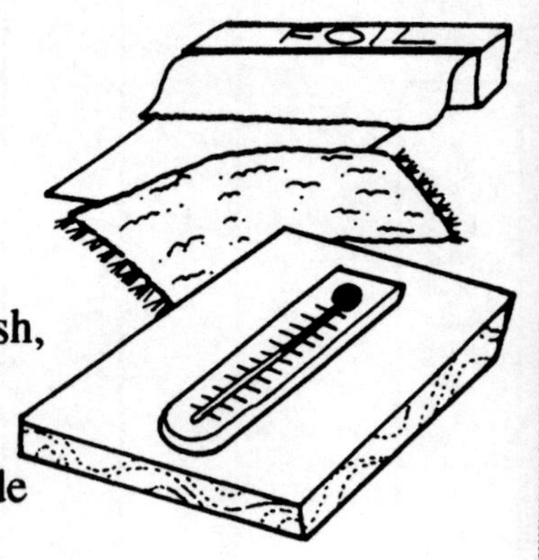

Materials

- Thermometers
- Sunshine
- Pencil
- Paper
- Piece of wood such as a wooden cutting board
- Variety of materials such as foil, paper, ceramic baking dish, towel, cardboard, etc.

Directions

1. Which of the materials collected do you think will provide the best insulation against heat?
2. Test your hypothesis by laying the thermometers out in direct sun on the piece of wood. You need the wood to act as an insulator because you want to measure only the amount of heat that comes through the test material. Cover the thermometers with the various materials you are testing.
4. Record your results. Which material provided the best insulation after five minutes? After ten minutes? After 30 minutes? Which material do you conclude would work the best on the outside of a spacecraft?

 © Edupress EP122

Heat Shields and Friction

Kitchen Quest

Search your kitchen for heat-resistant materials developed for use in space!

Materials

- Paper
- Pencil
- Resource book

Directions

1. Make a list of various materials around the house that are used at high temperatures.
2. Try to find the answer to this trivia question—what is the commercial name of *Pyroceram*, a material developed through space research? Is there an item made of this material on your list?

Corning Ware® ceramic cookware

Friction Demonstration

Demonstrate that when two surfaces rub together, they become heated because of the friction between them.

Materials

- Sandpaper
- Wood block

Directions

1. Rub the sandpaper over the wood very slowly for ten seconds. Feel the wood. How does it feel?
2. Try rubbing the sandpaper over the wood for ten seconds again—this time very fast. Touch the wood with your fingertips. How does it feel now?
3. Draw conclusions as to why people don't normally feel the heat of air friction.

© Edupress EP122

Weight and Space

Information

The apparent weight of an astronaut changes dramatically during his time in space! When an astronaut stands on Earth, the Earth pushes up on him with a force equal to his weight. This is called one *g-force*.

During a launch, the g-forces acting on the astronaut quickly build up to seven g's—increasing his weight seven times! This means that an astronaut weighing 150 pounds (63 kg) on Earth would weigh over 1,000 pounds (453.6 kg) during lift-off!

Riding on an orbiting spacecraft is similar to riding on a free-falling elevator. Because there is no force on his feet, an astronaut feels weightless. Weightlessness does not mean that there is no gravity. It means that the force of gravity is counterbalanced by the motion of the spacecraft.

At the end of the space flight, the g-forces acting on the astronaut may reach as high as 20 g's! The real weight of an astronaut does not actually change during a space flight—but it sure feels that way!

Project

Simulate the g-forces that act upon an astronaut during launch and reentry.

Materials

- Large plastic soda or juice bottle
- Pencil or wooden dowel
- Heavy rubber band
- Scissor
- Small heavy weight such as a lead sinker used by fishermen

Directions

1. Have an adult cut off the top of the bottle and cut two holes near the top for the pencil to pass through.
2. Tie one end of the rubber band to the weight and the other end to the pencil so that the weight hangs about midway in the bottle. The weight represents the astronaut and the bottle represents the spacecraft.
3. You can simulate the instant of launch by grasping the bottle with both hands and moving it sharply in an upwards direction. What do you think will happen to the length of the rubber band when you do?
4. "Launch" your rocket. Was your hypothesis correct? Now try moving the bottle sharply downward. What happens to the length of the rubber band now? Form your conclusions as to why this happens.

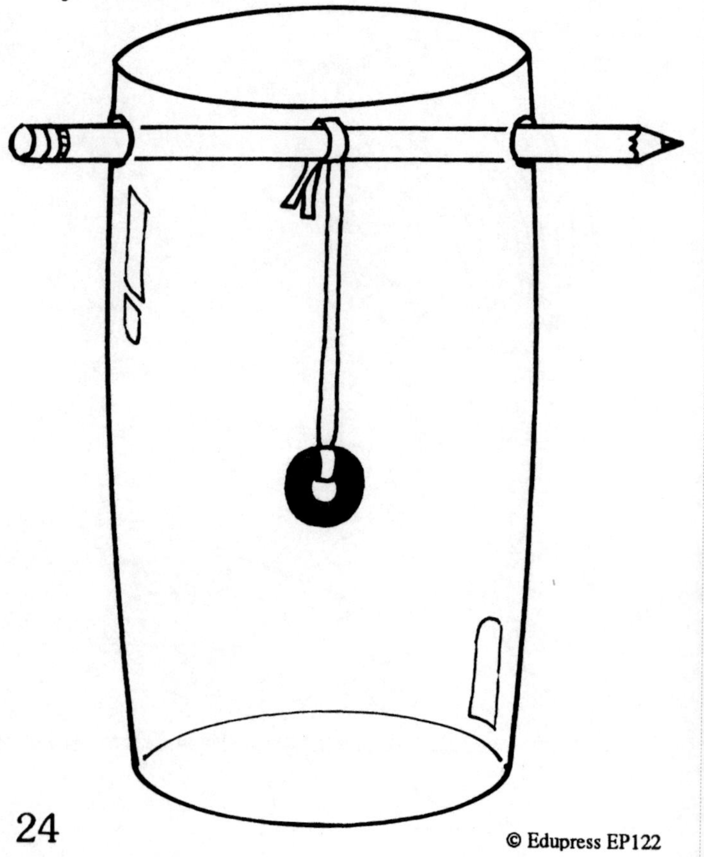

© Edupress EP122

Crewed Spacecraft

Information

After successfully launching several small satellites in the late 1950s, scientists and engineers in both the United States and the USSR were anxious to put men into space. The former Soviet Union was the first to launch a manned spacecraft—*Vostok 1* on April 12, 1961.

A similar program was underway in the United States. On May 5, of that same year, a Mercury spacecraft named *Freedom 7* carried the first U.S. astronaut into space. The series of twelve Gemini flights in 1965 and 1966 helped the United States develop the technology needed to land on the moon.

A year later, cosmonauts began to fly missions in Soyuz spacecraft that were designed to carry three people.

Space flights of the U.S. Apollo program began in 1968. On July 20 of the following year, astronauts landed the Apollo 11 lunar module, named *Eagle,* on the surface of the moon.

Project

Make a mobile of historic manned spacecraft.

Materials

- Pattern pages, following
- Colored poster board
- Crayons
- Scissors
- Hanger
- Pen or pencil
- Glue
- Holepunch
- Yarn or string
- Resource books

Directions

1. Reproduce the spacecraft patterns and the "Registration Card" master.
2. Research the spacecraft. Fill out a registration card for each vehicle.
3. Color, cut out, and mount the spacecraft and the registration cards onto poster board. Punch a hole in the top of each. Use yarn to tie the registration card to its matching spacecraft. Attach the spacecraft to the hanger to make a mobile.

 © Edupress EP122

Crewed Spacecraft Patterns

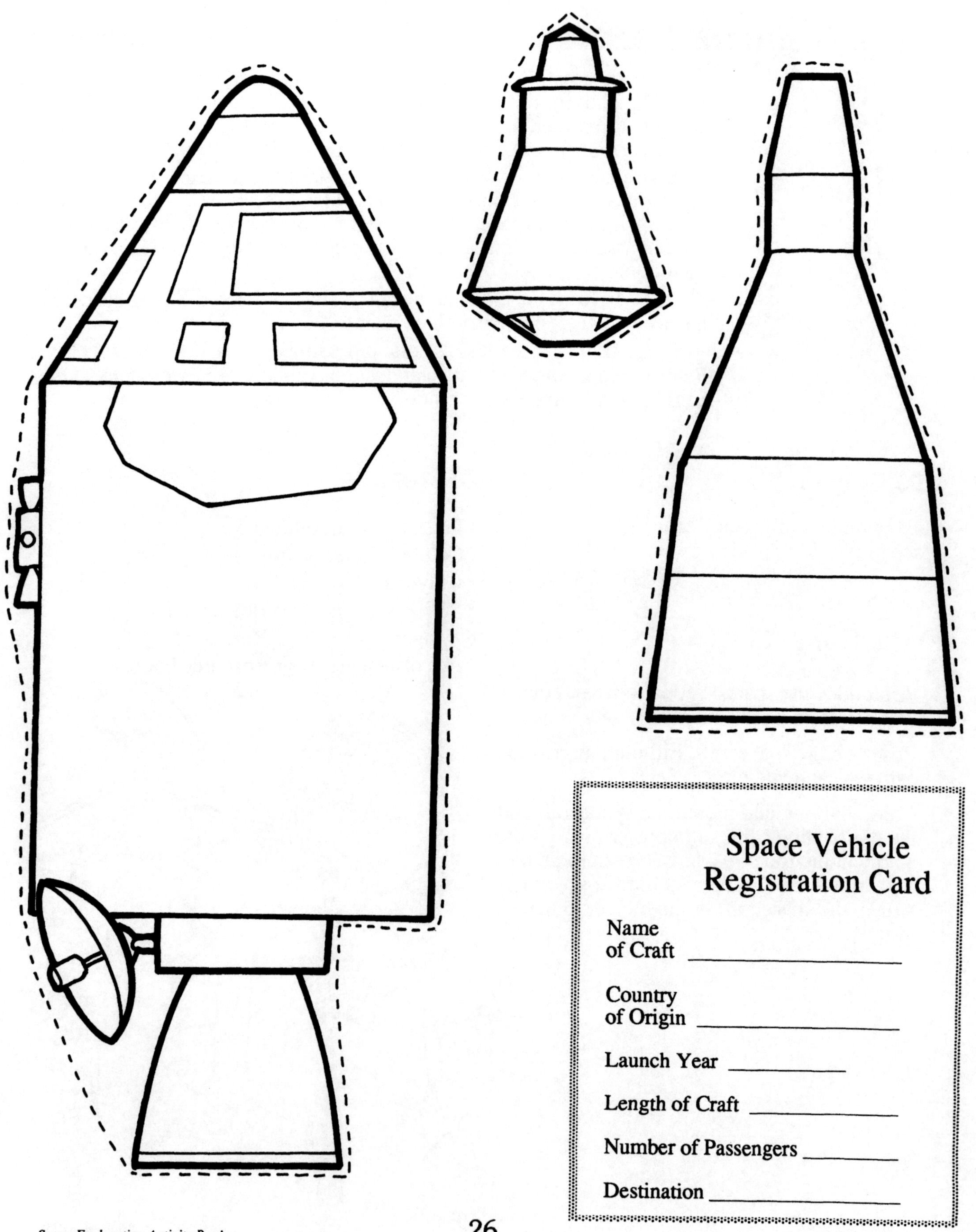

 © Edupress EP122

Crewed Spacecraft Patterns

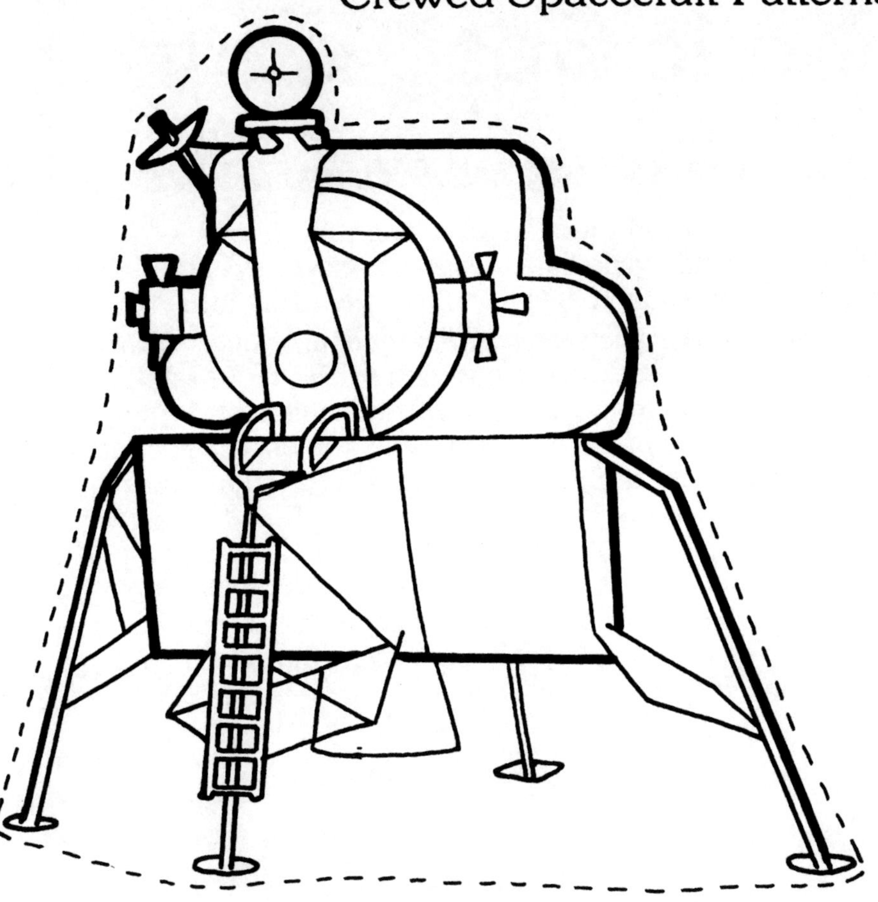

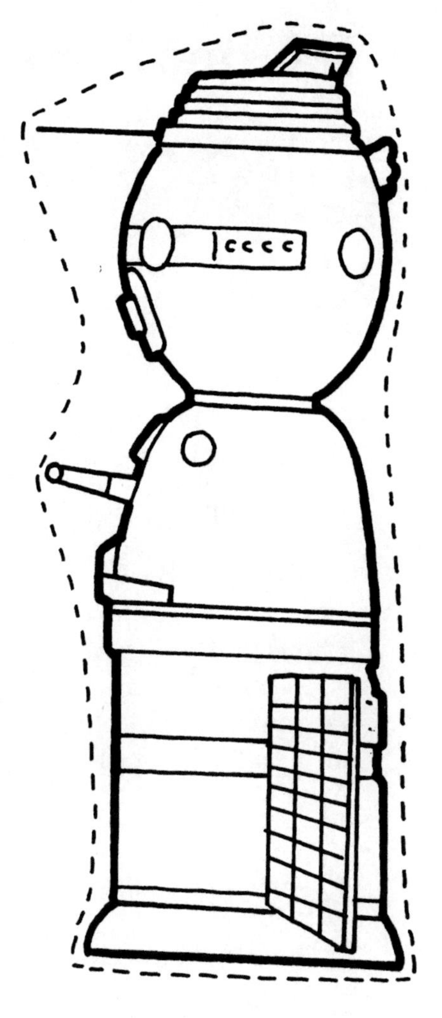

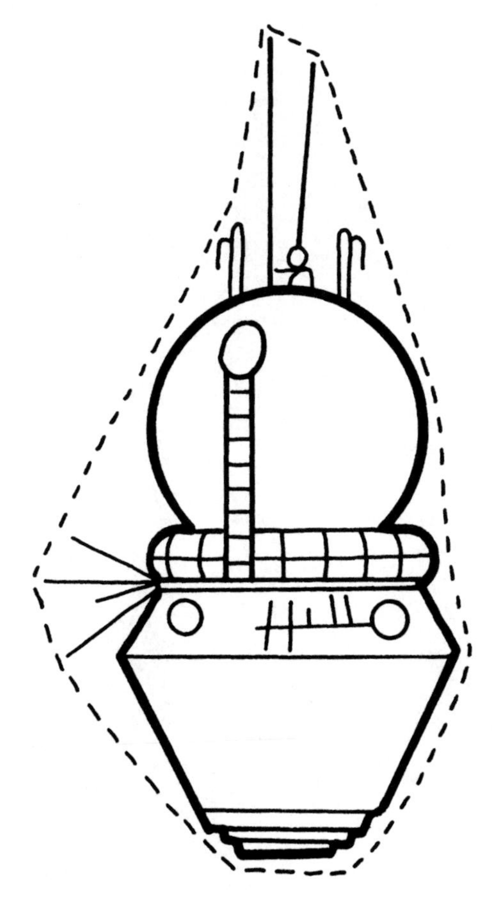

Space Vehicle Registration Card

Name of Craft ____________________

Country of Origin ____________________

Launch Year ____________

Length of Craft ______________

Number of Passengers __________

Destination ____________________

© Edupress EP122

Robotic Spacecraft

Information

Unmanned robots are launched from Earth with a mission—to explore space and send back data about our solar system. With each robotic spacecraft, or *probe,* that is launched, we learn more exciting information!

Some probes, such as *weather balloons* and *sounding rockets*, stay close to home to explore regions too near Earth for satellites to orbit. *Deep space probes* explore space between planets and do not reach a specific target in space. A *planetary probe's* instruments can collect a great deal of information about a planet, even from far away. Their on-board cameras send back amazing pictures! *Voyager 1* and *Voyager 2* sent back breathtaking photos of raging storms on the blue world of Neptune, the swirling red atmosphere of Jupiter, and the complex rings of Saturn.

Before the astronauts landed on the moon, probes named *Ranger, Surveyor,* and *Lunar Orbiter* sent back thousands of lunar pictures to determine the safest places to land.

Project

Form research groups to investigate different space probes.

Materials

- Research books
- Poster board
- Crayons, marking pens, colored pencils

Directions

1. Have each group choose a different space probe to research.
2. Make an exciting poster like the ones used at movie theaters. Include information about the probe's amazing mission in space. What does the robotic spacecraft look like? Where in the *universe* did it go on its travels? What information or pictures did it send back to Earth?

 © Edupress EP122

First Men on the Moon

Information

The first men on the moon were United States astronauts Neil A. Armstrong and Edwin E. Aldrin, Jr. On July 20, 1969, their Apollo 11 lunar module, *Eagle*, touched down on the moon. It landed on a rocky plain named the Sea of Tranquility. A third crew member, Michael Collins, stayed in lunar orbit during the landing. The spacecraft consisted of three *modules,* or detachable sections. The cone-shaped command module was the carrier for the astronauts and equipment. The service module held air and water supplies as well as housing the spacecraft's engines and fuel. The lunar module transported the astronauts to the surface of the moon.

When Neil Armstrong first stepped on the moon, he said, "That's one small step for (a) man, one giant leap for mankind." These first steps left deep footprints in the powdery lunar surface that are still there today! Armstrong posted a flag of the United States that was braced by wires so that it would appear to "wave" in the airless lunar atmosphere.

Project

Combine a crayon resist project and a writing activity to imagine what it might be like to be the first person on the moon.

Materials

- Large sheets of colored construction paper
- White paper
- Watercolor paints
- Crayon
- Paint brush
- Paste

Directions

1. Trace around your shoe with a crayon. Make the outline very thick and heavy. Add lines across the outline of the shoe to look like the treads on the bottom of a space boot. Paint over the boot design with a watercolor wash to create a crayon resist.
2. Imagine **you** are the first person on the moon. Write what you would say as you step onto the moon's surface—a place that no person has ever been before!
3. Mount your footprint and what you have written on the colored construction paper and display. Read what your friends have written!

 © Edupress EP122

Laboratories in Space

Information

Ordinary spacecraft are not really designed for experimental work. Both Russia and the United States have built special kinds of satellites to use as space laboratories. The first space laboratory was launched by Russia in 1971. Named *Salyut*, it was joined in space in 1973 by the American space laboratory, *Skylab*. European Space Agency scientists built the latest space laboratory, *Spacelab*. It is a reusable laboratory designed to fit in the payload bay of the space shuttle. The main part of *Spacelab* is a pressurized laboratory about the size of a school bus. There is also an unpressurized platform that carries the instruments that need to be exposed to space.

Space laboratories can be used not only to study the stars, our sun and other galaxies, but also the Earth and its atmosphere. Telescopes on board the orbiting labs are more effective than those on Earth as there is no atmosphere to dim the view. Physicists can study the laws of nature and biologists can learn more about the growth of plants and animals.

Project

Do an experiment that simulates a plant experiment done in space.

Directions

1. Duplicate the Growing Plants in Space chart.
2. Cut two long pieces of celery making sure they are the same size.
3. Have an adult cut a hole in the side of one of the containers the exact size of one of the pieces of celery. Insert the celery through the hole. Seal around the celery with clay so the water does not leak out. (You may have to place a stack of coins under the end of the celery to keep the container from tipping!)This celery is simulating a plant growing in space.
4. Stand the second piece of celery upright in the second container. This celery is simulating a plant growing on Earth.
5. Add several drops of food color to the water and pour the same amount into both containers.
6. Using the time intervals on the chart, measure how far the colored water has moved along the strip of celery at each checkpoint and record your observations.
7. After your tests are complete, draw a conclusion about growing plants in space. Write your ideas in the space provided.

Materials

- Growing Plants in Space, following
- Two identical disposable plastic containers
- Two pieces of celery, about ten inches (25.4 cm) long
- Water
- Food color
- Knife
- Ruler
- Modeling clay

© Edupress EP122

Growing Plants in Space

For plants to grow on Earth, dissolved minerals from the soil move upward through the roots and stems into the leaves. Try to answer this question— ***will plants grow faster in space where there is hardly any gravity to slow down the water as it moves along the stems?***

HYPOTHESIS

____

____

Plant Growing on Earth	
After 10 minutes	Water Measurement

Plant Growing on Earth	
After 20 minutes	Water Measurement

Plant Growing on Earth	
After 1 hour	Water Measurement

Plant Growing on Earth	
After 24 hours	Water Measurement

Plant Growing in Space	
After 10 minutes	Water Measurement

Plant Growing in Space	
After 20 minutes	Water Measurement

Plant Growing in Space	
After 1 hour	Water Measurement

Plant Growing in Space	
After 24 hours	Water Measurement

CONCLUSION

____

____

© Edupress EP122

Space Shuttle

Information

In 1981, astronauts traveled into space aboard *Columbia,* a new, reusable space craft called a space shuttle. The space shuttle is like three different space vehicles in one! It can launch like a rocket, orbit the Earth like a spacecraft, and land like an airplane.

Astronauts on the shuttle perform a variety of activities such as placing satellites into orbit or bringing them back to Earth. They can also repair damaged satellites, conduct experiments and study objects in space.

The two *solid rocket boosters* provide the thrust to launch the space shuttle. The *external fuel tank* carries more than 1½ million pounds (680,000 kg) of rocket fuel called propellent. The *orbiter* of the shuttle is the part that resembles an airplane and carries the astronauts and the cargo. Each orbiter is made to last for 100 flights.

Project

Build a model of the space shuttle.

Materials

- Paper towel tube
- Scissors
- Paint brush
- Potato chip can
- Tape
- Crayons
- Shuttle pattern, following
- Orange and white tempera paint
- Heavy paper or card stock
- Two to three-inch (5 to 7.54-cm) length of Velcro® fastening

Directions

1. Cut the paper towel tube in half vertically to make the solid rocket boosters.
2. Paint the rocket boosters white and the potato chip can orange to resemble the external fuel tank. Allow to dry.
3. Reproduce the shuttle pattern onto heavy paper or card stock.
4. Cut out the pattern, color, and fold as indicated.
5. Use Velcro® to attach the pieces of the shuttle together as shown.
6. Research how the space shuttle takes off, orbits, and lands. Use your model to demonstrate separating the boosters and tanks from the shuttle during launch, orbiting, maneuvering into position to land, and finally gliding to a landing. Just like the real shuttle, you can reassemble your shuttle and launch it over and over!

© Edupress EP122

Space Shuttle Pattern

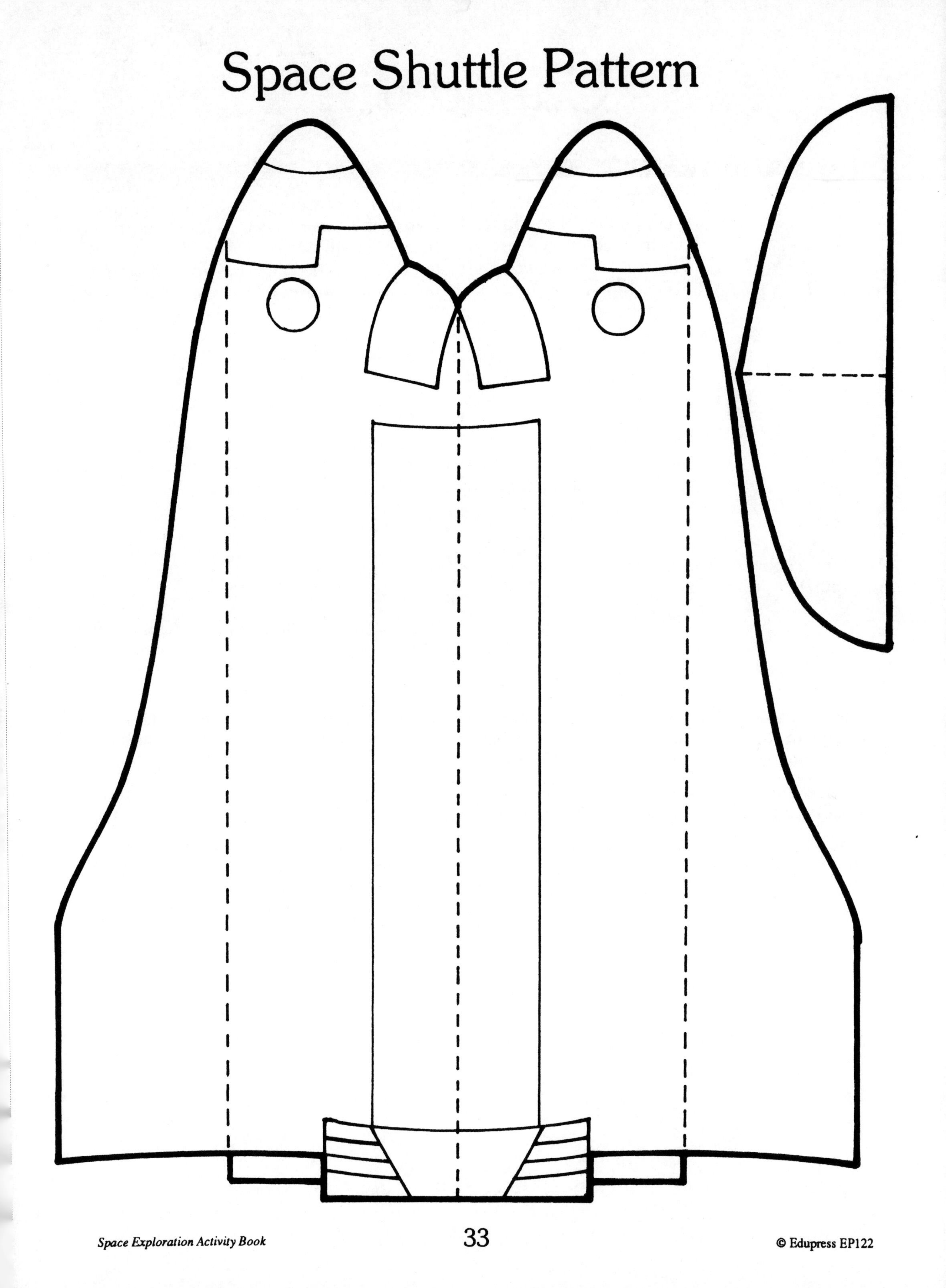

 © Edupress EP122

Canadarm

Information

The *Remote Manipulator System* was designed and built in Canada for use on the space shuttle. Known as the *Canadarm*, this robotic arm allows astronauts to release or retrieve satellites in space. The arm allows astronauts to work by remote control in the cargo bay from the safety of the flight deck. It has been used as a platform to hold astronauts making repairs to satellites outside the cargo bay. It has even dislodged ice from a shuttle vent!

The arm is operated by two hand controls on the flight decks. Video cameras located at the "elbow" and "wrist" of the arm assist the operator in the task. In space, the arm can lift a payload about the size and weight of a loaded school bus, but on Earth it cannot even support its own weight. Each arm is expected to last for 100 missions. Canadian engineers are designing and building new versions of the *Canadarm* for the International Space Station.

Project

Work as a cooperative team to manipulate a "robotic arm."

Materials

- Two players
- Blindfold
- Objects such as a paper cup, ruler, book, chalk, art eraser, marshmallow

Directions

1. The player who is going to be the robotic arm is blindfolded. The teacher writes the task on the board, such as, "Pick up the eraser from under the cup and put it on the book." The teacher then arranges the objects in front of the arm.
2. The second player is the astronaut. The astronaut cannot retrieve the object himself; he must operate the arm by remote control.
3. The astronaut must give verbal instructions to the arm to instruct it as to where the object is and how to pick it up. He or she can only use one-word commands such as "Forward," "Up," "Close," etc. See how many commands it takes for the arm to retrieve the object.
4. Try with another team. Who can complete their task using the fewest number of commands?

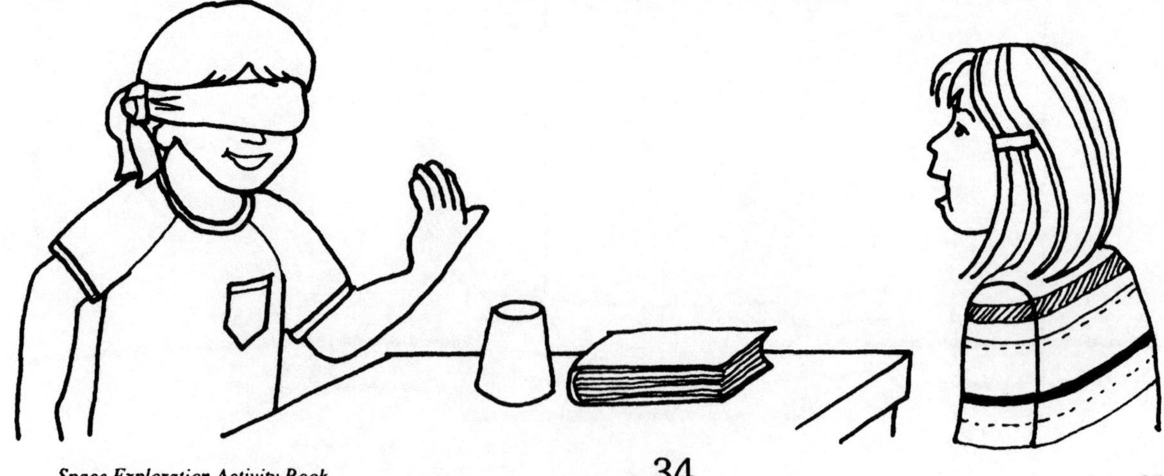

© Edupress EP122

Listening in Space

Information

Since 1960, astronomers have been aiming radio telescopes at all parts of the sky. They are listening for a signal from another planet. So far, all the radio waves that have been picked up are from natural sources only.

One of the problems facing researchers is how to identify a signal produced by an intelligent being on another planet. Scientists believe that they would be able to recognize such a signal by its regularity or pattern. It is thought that if such a signal should ever come, it would most likely be in the language of mathematics.

Another difficulty in communicating with another planet or distant star is the time it would take to send and receive messages. Radio waves travel at the speed of light which is 186,000 miles (299,000 km) per *second!* But even at that speed, it takes years for radio signals sent from Earth to reach even the closest stars!

Project

Create a code that could be read by beings on another planet.

Materials

- Paper and pencil
- Friend to decode your message

Directions

1. Draw a series of shapes to represent the numbers one to three, and the sign for plus and equals.
2. Write a set of simple problems using your symbols. Give it to a friend to decode.

1
2
3
4
=

© Edupress EP122

Exploring Mars

Information

On July 4, 1997, the Mars *Pathfinder* lander touched down on the surface of Mars. It had taken seven months to get from Earth to Mars!

Parachutes and airbags protected the lander as it dropped onto the planet's surface. The lander bounced at least 15 times without any of the airbags tearing. After the lander gently rolled to a stop, there was a tense moment as the airbags deflated and the lander opened. The project scientists were worried that the rover would get tangled in the airbags as it drove off the lander. They closed the lander, rolled up the airbags more tightly, and reopened the ramps—all from 111 million miles (178.6 million km) away! Then the small rover, named *Sojourner*, was sent out to explore the Martian surface and examine rock and soil samples.

The instruments on board the lander and on Sojourner sent back valuable information about the Martian soil, atmosphere, and weather for 83 days—three times longer than they were expected to!

Project

Design a lander that will protect an egg as it is dropped onto a planet's surface.

Materials

- Egg
- Various materials such as foam packing chips, bubble wrap, boxes, tape, balloons, etc.

Directions

1. Imagine you are a space engineer with the task of designing a lander that will safely get its cargo (the egg) onto a planet's surface after being launched from an orbiting spacecraft. Choose from the variety of materials to build your lander.
2. Find a safe place to drop your lander. How successful was your design? How would you change or improve a future design?

Fun to Do!

Read *The Story of A Little Rock* by Sue Kientz on the Internet at http://eis.jpl.nasa.gov/~skientz/little_rock/

 © Edupress EP122

The Hubble Telescope

Information

The Hubble Telescope is one of four space telescopes known as the Great Observatories. The Hubble space telescope was launched in April of 1990. This powerful telescope, about the length of a large school bus, can take pictures of large sections of the sky and extremely detailed pictures of nearer objects such as the planets in our solar system.

When scientists studied the first blurred images sent back by Hubble, it was discovered that the telescope's main mirror had a flaw. To correct this problem, seven astronauts aboard the space shuttle *Endeavour* made a mission in 1993 to replace some of the scientific instruments.

The repaired telescope works perfectly. It recently sent back images of what scientists think may be the first picture ever of a planet outside our solar system! Named TMR-1C, this giant is two or three times the size of our largest planet, Jupiter.

Project

Make a replica of an image sent back to Earth by the Hubble Telescope.

Materials

- Green, purple, red, and black tempera paint
- Toothbrush
- White construction paper
- Scrap of colored construction paper

Directions

1. Tear the scrap of colored construction paper into a shape similar to the one shown below to make a stencil.
2. Lay the cut stencil on the white construction paper. Dip the bristles of the toothbrush into the black paint. Hold the toothbrush over the pattern and gently shake it. The paint should spatter onto the construction paper. Use a lot of black to cover the construction paper, then add the other colors. The heaviest concentration of the bright colors should be in the center of the stencil. Allow to dry.

 © Edupress EP122

Mission to Planet Earth

Information

Mission to Planet Earth is a program designed to study our most important planet in the solar system—our home planet Earth!

Scientists want to measure chemicals in the air and oceans, make maps of plant communities and take temperature readings to help them understand our world better. To collect this data, a family of satellites called the *Earth Observing System* (EOS) was developed and launched beginning in 1997.

By monitoring such things as temperature, ocean currents, emission of greenhouse gasses, the extent of pollution, the water cycle and the rate of rain forest destruction scientists can evaluate the impact human beings are having on the environment.

Mission to Planet Earth will help scientists predict weather patterns that lead to droughts and floods, global warming, and severe storms.

Project

Write a poem that expresses our care for our home planet.

Directions

1. Draw a picture of the Earth in space using the colored pencils and pressing very lightly to create a pale background picture.
2. On a piece of scrap paper, write a poem about the Earth and our need to care for our home.
3. Copy the poem onto the picture of the Earth.
4. Mount on colored construction paper.

Materials

- Paper
- Colored pencils
- Pen
- Colored construction paper
- Glue

© Edupress EP122

Satellites

Information

Satellites come in many different shapes and sizes and serve a variety of functions. *Communications satellites* make it possible to send television programs, radio messages, and phone calls around the world. *Navigation satellites* help airplane pilots and sailors find their position even at night or in bad weather. *Scientific satellites* carry a variety of instruments around the Earth that explore the Earth's atmosphere and take measurements of radiation levels and magnetic fields.

Meteorological satellites, or weather satellites, help scientists forecast the weather and learn about the Earth's atmosphere and how weather is made. These satellites take pictures of the Earth's surface showing areas of ice and snow and how the clouds are moving. By studying these pictures, meteorologists can discover if a storm is forming. Other instruments measure heat coming from the Earth and the clouds. Scientists use this information to better study how weather is related to the heating and cooling of the Earth.

Project

Give a weather report to the class using a "satellite photo."

Materials

- Butcher paper
- Crayons
- Tape
- Cotton balls
- Glue
- Access to television news program

Directions

1. Watch the portion of a television newscast that gives the weather forecast. Note the satellite picture that shows the clouds in your part of the country. Listen to the weather forecaster talk about the movement of these clouds and how it will affect your weather.
2. Draw a large outline map of the states around where you live on butcher paper. Use cotton balls to add some clouds to resemble the satellite picture you saw on the news.
3. Post your satellite picture and imagine you are a meteorologist giving a weather report on television. Explain to your audience what a satellite picture is and where it came from. Explain how meteorological satellites help scientists such as yourself predict the weather.

 © Edupress EP122

Space Stations

Information

A space station is a satellite designed so people can live and work in space for long periods of time. The former Soviet Union launched the space station *Mir* in February 1986. *Mir* could be rearranged into different configurations, and other spacecraft were able to *rendezvous* with *Mir* to change crews, deliver supplies, and take data back to Earth. The former Soviet Union worked with the US to create the plans for the *International Space Station*.

In 1998, 16 nations worked together to construct the *ISS*. The Station includes docking platforms, living quarters, supply canisters, and laboratory *modules* attached to a scaffold and connected by tunnels called *nodes*. In 1998, the first two modules of the *International Space Station* were launched and joined together in orbit. The first crew arrived in 2000. The Station is a giant laboratory for learning how to live and work in space. This information helps astronauts prepare for future trips to the moon, Mars, and beyond!

Project

Color a diagram of the largest, most complex structure that has ever been placed in orbit—the *International Space Station*.

Materials

- *International Space Station* diagram, following
- Pencil
- Paper
- Crayons or colored pencils

Directions

1. Reproduce the *International Space Station* diagram. Use the guide to help you color the different sections.
2. After becoming familiar with some of the necessary components on a space station, design and label your own space station of the future!

The preliminary plans for the ISS included the parts labeled on the diagram to the right.

1. **Solar Arrays**—provide the power needed for the space station. Color them blue.
2. **The Canadarm**—a robotic tool used to help build the station. It moves along rails covering the length of the station. Color it red.
3. **Habitation Module**—where the astronauts live. Color it green.
4. **Service Module**—carries the life support for the entire station. Color it orange.
5. ***Soyuz* Spacecraft**—This spacecraft is docked to the station. Its purpose is to serve as a rescue vehicle! Color it purple.
6. **Laboratories**—Experiments are sent to the station in racks that look like filing cabinets! Color them pink.
7. **Main Truss**—the central scaffold. Color it gray.
8. **Thermal Radiators**—Color them black.

 © Edupress EP122

The International Space Station—2003

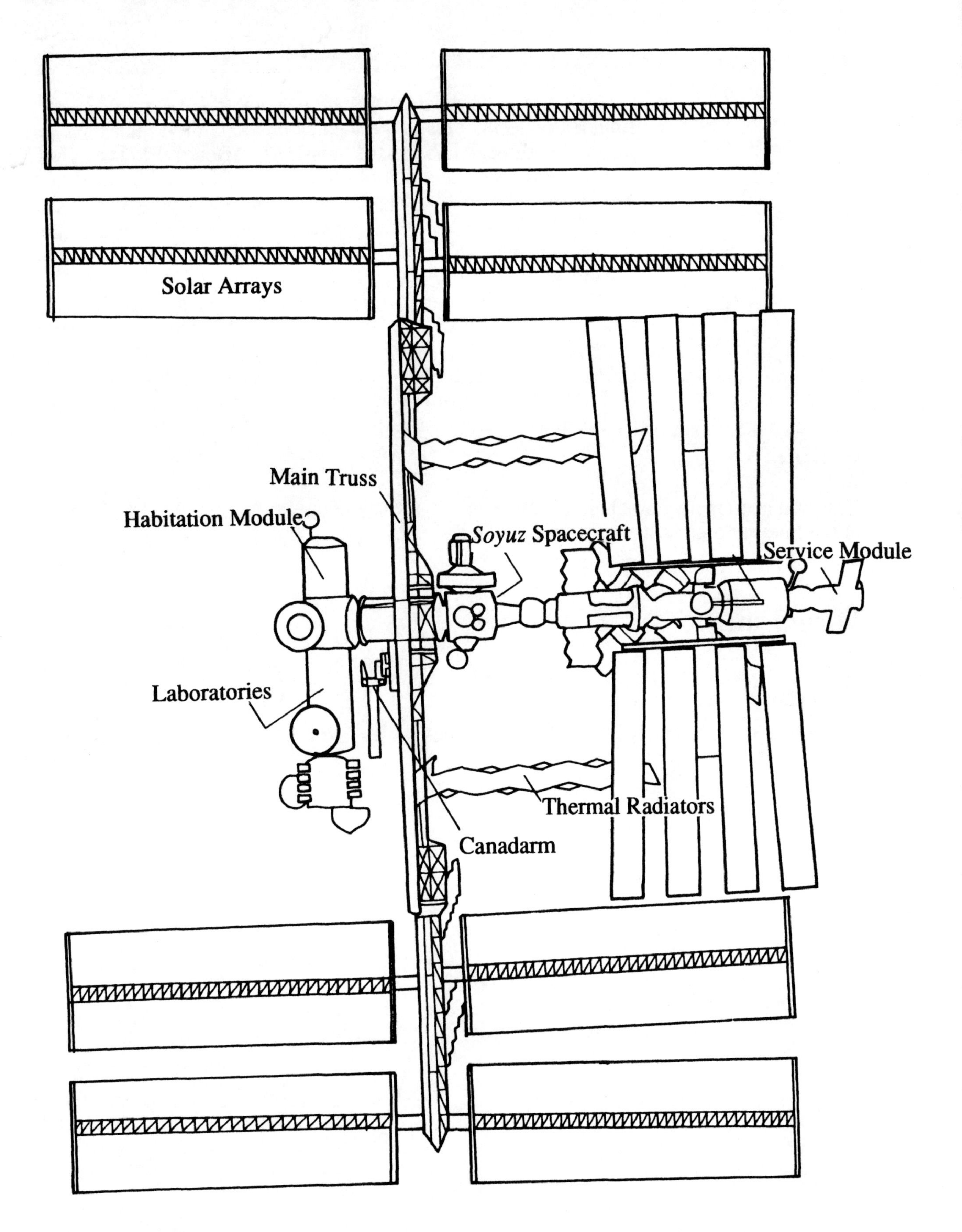

 © Edupress EP122

Space Agencies of the World

Information

The *National Aeronautics and Space Administration,* or *NASA,* was founded in 1958. This is the agency of the United States Government that plans and directs U.S. space missions. After the breakup of the Soviet Union the *Russian Space Agency, RKA,* was formed. It uses the technology and launch sites that belonged to the former Soviet Space Program.

The *National Space Development Agency of Japan*, known as *NASDA* was established in 1969. *Tanegashima Space Center* is the largest of Japan's launch facilities.

Four years later, ten European countries met to form *ESA*, the *European Space Agency*. It became official in 1980 and now has fourteen members.

The *Canada Space Agency* was officially established in 1989 by an act of Parliament. Canada's experience in space actually began 27 years earlier with the launching of the *Alouette 1* research satellite. Canada became the first country in the world after Russia and the U.S. to build its own satellite.

Project

Research and create a brochure for one of the international space agencies.

Materials

- White paper
- Colored pencils
- Resource books

Directions

1. Choose one of the space agencies listed in the above text to research.
2. Accordion-fold the paper into thirds. On the first page, write the name of the space agency, draw and color its logo.
3. On the remaining pages—
   - draw a picture of a spacecraft pilot wearing the uniform of that country.
   - write a brief history of the agency.
   - list some of the agency's accomplishments in space.
   - write a brief biography of a few of the astronauts or cosmonauts.

© Edupress EP122

Industry in Space

Information

The airless environment and *microgravity* in space make it an ideal place for some industries to produce better products. Perfectly formed ball bearings, crystals, and silicon chips used in computers are a few examples.

A major industry in space will be the production of solar energy. In space there is no atmosphere to block the sun's radiation. Vast groupings of solar cells will convert sunlight to microwaves that can pass through Earth's clouds. Receiving stations on the surface of the Earth will be able to convert the microwaves to electricity.

Mining will be another important space industry of the future. Asteroids are rich sources of many metals, including gold and platinum. These materials would be mined and brought back to Earth. A mine built on the moon could yield the aluminum, titanium, and iron needed to build solar power stations in space.

Project

Form cooperative groups to create a newspaper page featuring information on space industries.

Materials

- Paper
- Pencil

Directions

1. Create newspaper page. Make one of the sections the Want Ads. What kinds of jobs will there be in space in the future? What qualities will people need to work and live in space? Where will the engineers and workers and their families live? What facilities will be provided?
2. Design advertisements on the page featuring products made in space. One might be an ad for a power company that sells solar power from space. What are the advantages of the products featured in your ads?
3. Don't forget to include a travel section with information on space travel for people on vacation!

 © Edupress EP122

Energy in Space

Information

All spacecraft require power to keep their instruments functioning and to maintain radio communication with Earth. Manned spacecraft have even greater energy needs. The astronauts on board must have power to run air conditioners, stoves, and refrigerators. They need lights during the time they are in the Earth's shadow.

The energy the spacecraft needs comes from the sun. Wafer-like silicon cells, called solar cells, convert the sun's energy into electricity when exposed to sunlight. Batteries provide electricity when the spacecraft is in the Earth's shadow. These batteries are, in turn, recharged by the solar cells.

Solar energy can be used to heat air and water on a spacecraft. In fact, the heat produced by the sun can raise temperatures so high that it becomes a problem. The challenge then is to get *rid* of the excess solar energy!

Project

Use the energy from the sun to make a class treat!

Materials

- Large jar with a tight-fitting lid
- Cold water
- Two raspberry herb tea bags per quart (liter) of water
- One orange
- Sugar (optional)
- Cups

Directions

1. Fill the jar with water. Cut the unpeeled orange into small pieces. Add the orange and the tea bags to the jar. Tighten the lid closed.
2. Place the jar in full sunlight. The jar must be located in the sun's direct rays. It will take about two hours to brew a jar of sun tea.

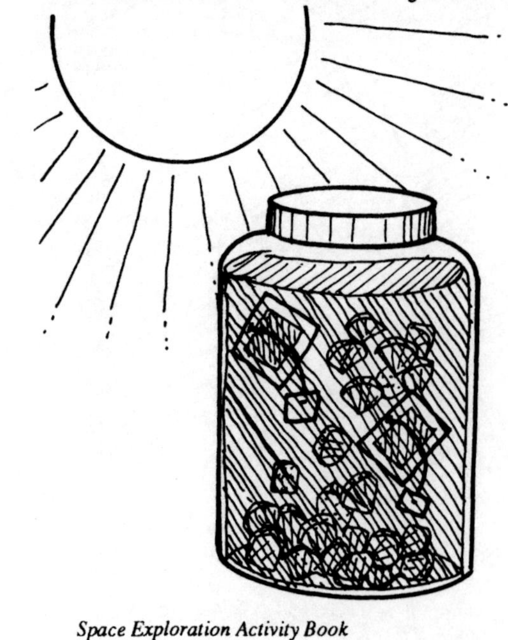

© Edupress EP122

Day in Space

Project

Turn the classroom into a spacecraft and experience a day in space.

Materials

• See individual stations

Directions

1. Send a note home in advance to share with parents the details of your classroom's trip into space. On the day of the "launch" have children bring in a lunch packed with items that could be eaten in space. (See *Eating in Space,* page 11.) Children can wear sweat suits to school and bring in their backpack, boots, and gloves to complete their space suit.
2. Using the activities in this book as a guide, set up the stations in the classroom. Some suggestions for centers are on the following page. Rotate throughout the day so everyone has a complete adventure in space.

The activities on this page should be done before launch day.

Space Helmets

Materials

• Poster board
• Tape
• Narrow flexible tubing OR construction paper
• Backpack
• Scissors

Directions

1. Cut an opening in the poster board as shown. Bend into a cylinder to fit over head and tape.
2. Cut a three-foot (91-cm) piece of tubing. Cut a hole in the poster board to fit the tubing **or** accordion-pleat long strips of construction paper to make a three-foot (91-cm) piece. Tape to the helmet to make the air tube.
3. To wear, put on the helmet and backpack. Tuck the end of the air hose into the life support system backpack.

Earth Mural

Materials

• Butcher paper
• Tempera paints

Directions

1. Paint a mural of Earth from space to hang over the windows of the classroom.

Patch

Materials

• Construction paper
• Crayons
• Scissors
• Safety pins

Directions

1. Create a custom classroom patch to wear on your space suits. Color, cut out and pin on.

 © Edupress EP122

Earth Mural

Get ready for a fun day in space!

Materials
- Recording of an actual launch
- Fabric or felt strips (optional)

Directions
1. Prepare for launch by sitting in the cockpit seats and fastening either imaginary seatbelts or the fabric strips.
2. Imagine what an astronaut is feeling as he or she is hurtled out into space!

From Launch to Lunch!

Enjoy the lunch brought from Earth as well as other food-related activities!

Materials
- See activities on pages 16-17

Directions

Set up a station where astronauts can mix and taste dehydrated foods.

Journal Writing

Allow time for the astronauts to record the day's events in their space journal.

Materials
- Journal, pages 20-21
- Pencils
- Stapler
- Yarn

Directions
1. Assemble the journals and tie a pencil to each one so they won't float away!
2. Have the journals ready to distribute before launch.

Exercise

Astronauts must exercise at least 30 minutes a day to keep their bones from weakening.

Materials
- List of exercises on page 11

Directions
1. Spend some time keeping your bones in good condition!

Relaxation

Pretend you are playing a game of kickball on an orbiting space station!

Materials
- Balloons
- Two or more players

Directions

Blow up the balloons and tie them tightly closed. Lie on the floor and pretend the floor is the wall of the spaceship. Kick the balloon "ball" to your crewmate. Try to keep it from hitting the wall of the spacecraft!

Working in Space

Materials
- Helmets, page 45
- Gloves
- Backpack
- Variety of small materials to manipulate such as nuts and bolts

Directions

Set up this station outside the classroom, either in the hallway or outdoors. Astronauts can put on their helmets and gloves and go to this station to "repair the satellite." They must keep their gloves on as they try to assemble and disassemble the nuts and bolts.

Experiments in Space

Make one station a laboratory where students can carry out experiments in the book.

Materials
- See individual experiments for material needed
- Paper
- Clipboard
- Pencil
- Yarn

Directions
1. Attach the pencil to the clipboard. If possible, hang the clipboard in your "lab."
2. Have the space scientists carry out an experiment and record their results.

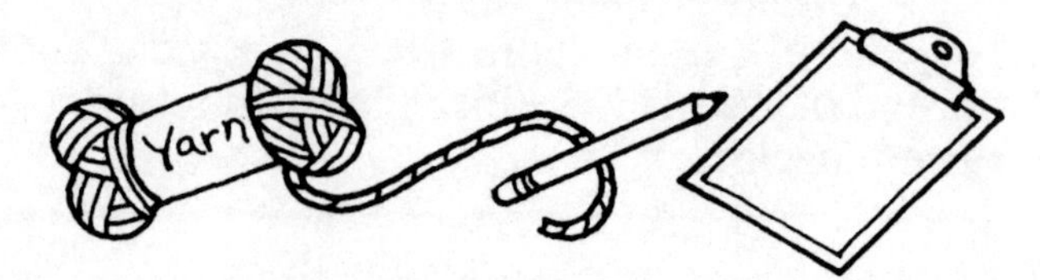

© Edupress EP122

Space Race

Project

Have a "Space Race" using the vocabulary words from the glossary.

Materials

- Black or dark blue butcher paper
- Glossary, page 4
- Rocket pattern, below
- Star stickers
- Tape
- Scissors
- Marking pen

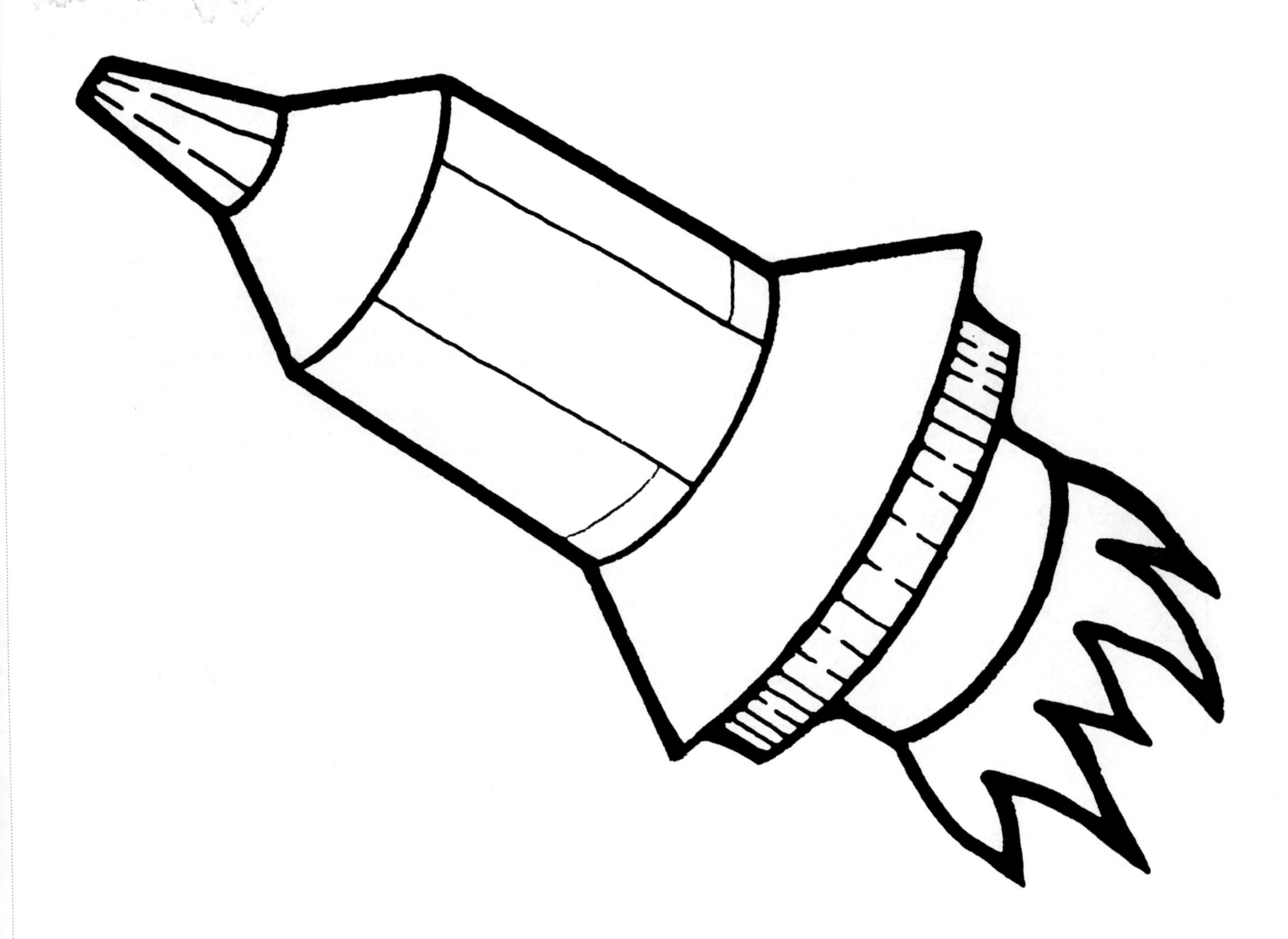

Directions

1. Cover the bulletin board with butcher paper. Reproduce the rocket pattern and cut out. Write the vocabulary words on the rockets and tape to the butcher paper. Decorate with star stickers.
2. On the day of the "Space Race," have a spelling bee using the vocabulary words and definitions instead of spelling words. The astronauts start out in a line on Earth. For every right answer, they get to take a "giant step for a man." The astronaut who has travelled the farthest at the end of the game is the winner of the space race!

© Edupress EP122

World Wide Web

Are you anxious to take off on space adventures of your own? Try a flight on the World Wide Web. The web pages below were active at publication date but their continued presence is not guaranteed. There's a whole world of space discoveries waiting for you.

Address	Content
www.msfx.nasa.gov	*NASA Home Page*—up-to-date NASA information and links to lots of other great websites.
http://k12unix.larc.nasa.gov/projects/earthkam	*EarthKAM*—put a camera on the shuttle, point it at Earth and use it to investigate Earth from space.
http://www.spacecamp.com/	*U.S. Space Camp*—besides information on Astronaut and Aviation Challenge programs, you'll find the U.S. astronaut Hall of Fame and links to the U.S. Space and Rocket Center.
http://www.space.gc.ca	*Canadian Space Agency*—in English or French, explore the latest information on Canada's Space Program. Visit the Kool Zone for video clips.
www.ksc.nasa.gov	*Kennedy Space Center Home Page*—information on the Space Shuttle and its missions, Space Flight Archives, Shuttle Countdown.

 © Edupress EP122